# International Science

## Workbook 1

# International Science

Workbook 1

Karen Morrison

HODDER
EDUCATION
AN HACHETTE UK COMPANY

The author and publisher would like to thank the following for permission to reprint photos in this book:

**p.57** Mike van der Wolk; **p.67** Mike van der Wolk

Every effort has been made to trace all copyright holders, but if any have been inadvertently overlooked the Publishers will be pleased to make the necessary arrangements at the first opportunity.

Hachette Livre UK's policy is to use papers that are natural, renewable and recyclable products and made from wood grown in sustainable forests. The logging and manufacturing processes are expected to conform to the environmental regulations of the country of origin.

Orders: please contact Bookpoint Ltd, 130 Milton Park, Abingdon, Oxon OX14 4SB. Telephone: (44) 01235 827720. Fax: (44) 01235 400454. Lines are open 9.00–5.00, Monday to Saturday, with a 24-hour message answering service. Visit our website at www.hoddereducation.co.uk.

Impression number    9
Year                 2017

Cover photo © Alan Schein Photography/Corbis
Illustrations by Barking Dog Art and Richard Duszczak
Typeset in 12.5/15.5pt Garamond by Charon Tec Ltd (a Macmillan Company)
Printed and bound by CPI Group (UK) Ltd, Croydon, CR0 4YY

A catalogue record for this title is available from the British Library

ISBN 978 0 340 96600 6

# Contents

# Doing science

**Activity 1** **Safety rules**                    Date: _____

**1** Look at the drawing of a science classroom carefully.

**2** Draw a red circle around each danger that you can see in this classroom.

**3** Choose three of the dangers you have circled. Write down a safety rule for each one to make the classroom safer.

Rule 1: _____

_____

Rule 2: _____

_____

Rule 3: _____

_____

## Activity 2  Science words

Date: _____

> survey    opinion    fact    dependent variable    conclusion    fair test
> independent variable    discrete data    measuring    continuous data

**1** Write the ten terms from the box above in alphabetical order in the first column of this table.

| Term | Term used in a sentence |
|------|--------------------------|
|      |                          |
|      |                          |
|      |                          |
|      |                          |
|      |                          |
|      |                          |
|      |                          |
|      |                          |
|      |                          |
|      |                          |

**2** Use the glossary on pages 175 to 178 of your coursebook to find the terms and their meanings.

**3** Write a sentence next to each term in the table to show how the term is used in science.

# Activity 3  Labelling diagrams

Date:_____

**1** Pretend you are a science teacher, marking homework. Put a red circle around each mistake in this pupil's labelled drawing.

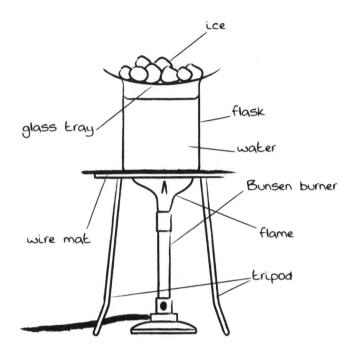

**2** Label this diagram correctly. The words you need are in the box below the diagram.

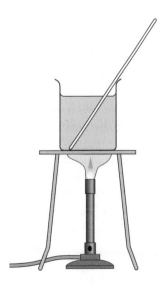

beaker   flame   stirring spoon   Bunsen burner   wire mat   salt and water solution

## Activity 4 — Working with graphs

Date:_____

### Which type of graph is best for each investigation?

A
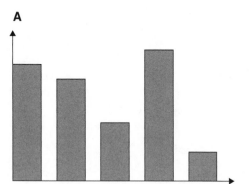

B
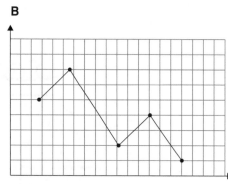

C
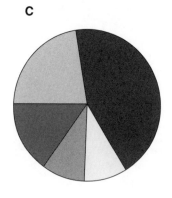

**1** You measure the temperature in the classroom at the end of each lesson to see how it changes during the day.

  **a)** Which type of graph would you use to show your results? _____

  **b)** Write a title for the graph you have chosen.

  _____

  **c)** Explain why you chose that type of graph.

  _____

**2** You investigate which colour of car is most popular in your area by counting all the white, red, yellow, black and silver cars that pass your school in an hour.

  **a)** Which type of graph would you use to show your results? _____

  **b)** Write a title for the graph you have chosen.

  _____

  **c)** Explain why you chose that type of graph.

  _____

**3** You do a survey of 100 people to find out which language they speak at home. You find there are five languages and you find out how many people speak each language.

  **a)** Which type of graph would you use to show your results? _____

  **b)** Write a title for the graph you have chosen.

  _____

  **c)** Explain why you chose that type of graph.

  _____

## Drawing your own graph

Sindi and Lucas did an investigation to measure the temperature changes in hot and cold water. The diagram shows what they did.

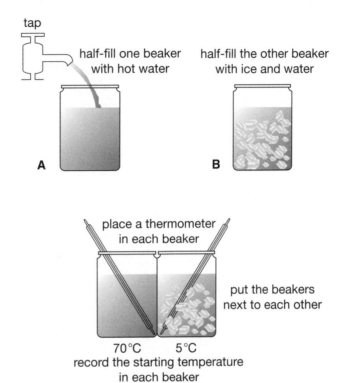

The table shows the changes in temperature that they recorded.

| Time after start (minutes) | Temperature in beaker A (°C) | Temperature in beaker B (°C) |
|---|---|---|
| 0 | 70 | 5 |
| 2 | 65 | 8 |
| 4 | 58 | 11 |
| 6 | 50 | 15 |
| 8 | 45 | 16 |
| 10 | 40 | 19 |
| 12 | 38 | 20 |
| 14 | 34 | 21 |
| 16 | 30 | 22 |
| 18 | 28 | 24 |
| 20 | 25 | 25 |

Draw a graph to show these results – use the graph paper given below.

Use a different colour for each beaker.

Give your graph a title.

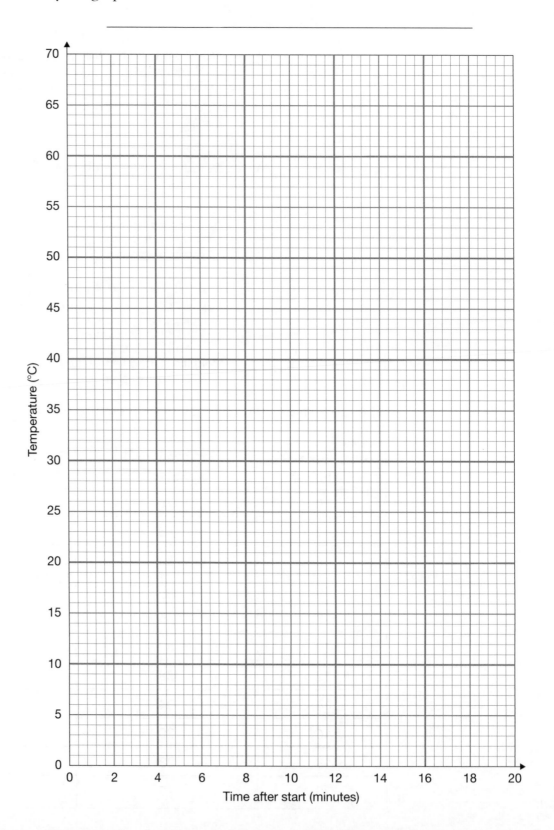

# Chapter 2 — Characteristics of living things

## Activity 1  Words and their meanings

Date: _____

There are seven characteristics of living things.

Draw lines to match each characteristic to its correct definition.

If you cannot remember what each word means, look at the glossary on pages 175 to 178 of your coursebook.

| Characteristic | Definition |
|---|---|
| excretion | responding to what is happening in the world around them |
| movement | feeding and taking in food |
| respiration | changing position and location by moving body parts or moving from place to place |
| nutrition | making more living things like themselves |
| reproduction | getting bigger |
| sensitivity | getting rid of waste products |
| growth | using oxygen they breathe in to make their bodies work |

## Activity 2  Observing animals

Date:_____

You are going to observe three animals in your local environment.
You will need to choose a small animal, a medium-sized animal and a large animal.

You will observe each animal for ten minutes.
Complete the tables below to show what you observed in ten minutes.

### Small animal

Type of animal: _____

|  | Yes or no | How often? |
|---|---|---|
| I saw the animal move |  |  |
| I saw the animal breathe |  |  |
| I saw the animal eat |  |  |
| I saw the animal excrete |  |  |

### Medium-sized animal

Type of animal: _____

|  | Yes or no | How often? |
|---|---|---|
| I saw the animal move |  |  |
| I saw the animal breathe |  |  |
| I saw the animal eat |  |  |
| I saw the animal excrete |  |  |

### Large animal

Type of animal: _____

|  | Yes or no | How often? |
|---|---|---|
| I saw the animal move |  |  |
| I saw the animal breathe |  |  |
| I saw the animal eat |  |  |
| I saw the animal excrete |  |  |

What differences did you notice between these animals?

_____

_____

_____

_____

**Activity 3** **Observing plants**                    Date:_____

**1** Label this diagram to show what happens when a bean seed germinates and starts to grow into a bean plant.

  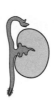 

**2** Find seeds from five different plants (A–E), and draw an example of each in the boxes below.

| Seed A | Seed B | Seed C | Seed D | Seed E |
|---|---|---|---|---|
| | | | | |
| **Description** | **Description** | **Description** | **Description** | **Description** |
| | | | | |

**3** Underneath each drawing, write a description of each seed. Use these words to help you.

| Size | large | medium | small | very small |
|---|---|---|---|---|
| **Shape** | round | flat | long | short |
| **Colour** | brown | white | red | green |
| **Texture (feel)** | rough | smooth | hairy | feathery |

**Activity 4** **Good conditions for living things**  Date:_____

Living organisms can only live if they have the right conditions.

Name each of the living organisms shown in the table below.
Then complete the table by filling in three things that each organism needs in order to live.

| | Name of organism | Three things it needs to live |
|---|---|---|
| | | |
| | | |
| | | |
| | | |
| | | |
| | | |
| | | |

# Chapter 3 Cells and organ systems

## Activity 1 The microscope

Date:_____

**1** Label the parts of the microscope shown in the diagram.

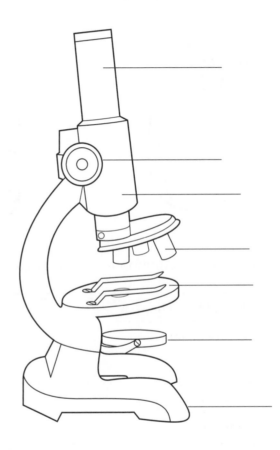

**2** Now write the name of each part next to its description in the table.

| Part | Description |
|------|-------------|
| | the platform that the microscope stands on |
| | the lens you look through to see the thing you are studying |
| | the piece that holds the object you are studying |
| | the lens that is near to the thing you are studying |
| | the knob you turn to focus the image of what you are studying |
| | the tube between the two lenses |
| | shines light onto the thing you are studying |

**Activity 2** **Plant and animal cells**     Date: _____

**1** Use the terms in the box to correctly label the parts of the plant cell and the animal cell shown in the diagram.

| cytoplasm | cell wall | vacuole | nucleus | cell membrane | chloroplasts |

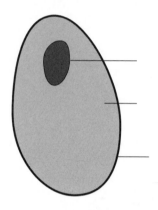 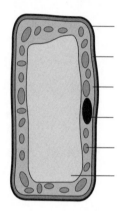

**2** Write a definition to say what each part of the cell is and what it does.

| Part of the cell | Definition |
|---|---|
| cell membrane | |
| cell wall | |
| nucleus | |
| cytoplasm | |
| vacuole | |
| chloroplasts | |

**Activity 3** ## Specialised cells

Date:_____

1 In the table below, draw each type of specialised animal cell listed.

2 Fill in what each cell does (its function) in the body.

| Type of cell | What it does |
| --- | --- |
| sperm cell | |
| egg cell | |
| red blood cell | |
| white blood cell | |
| muscle cell | |
| nerve cell | |
| goblet cell | |

3 Give three examples of specialised cells found in plants and say what each one does.

    **a)** Type of cell: _____

      What it does: _____

    **b)** Type of cell: _____

      What it does: _____

    **c)** Type of cell: _____

      What it does: _____

## **Activity 4** **Body parts**

Date: _____

**1** Copy the body parts shown in the box in the correct positions on the human body. If you are a girl, fill in the female diagram. If you are a boy, fill in the male diagram.

**2** Label each part that you have filled in on your diagram.

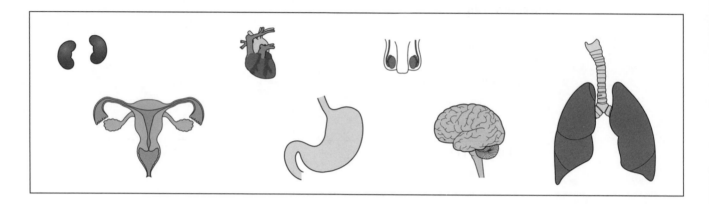

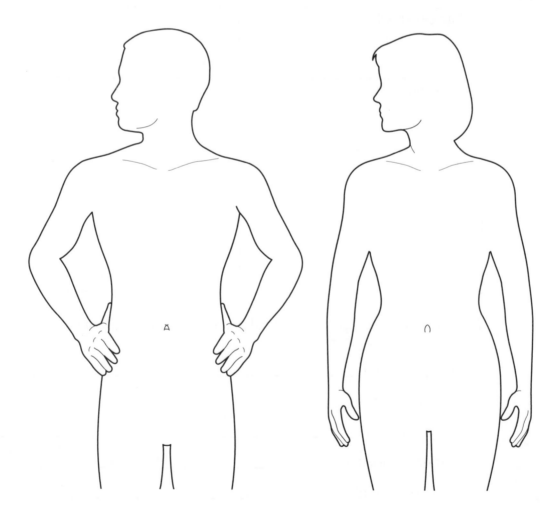

# Activity 5   Organs and organ systems

Date: _____

**1** Study the diagrams of different organs found in the human body.
Write the name of each organ next to it.

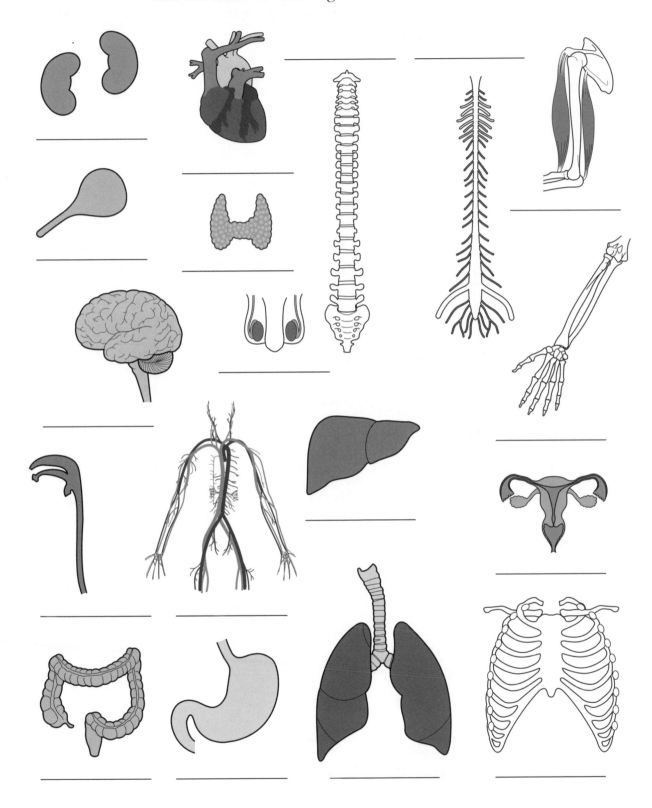

**2** Write the name of each organ from question 1 in the correct place in the table below, to show which organ system it belongs to.

| Organ system | Organs that form part of this system | The main function of this organ system |
| --- | --- | --- |
| digestive system | | |
| excretory system | | |
| respiratory system | | |
| circulatory system | | |
| endocrine system | | |
| reproductive system | | |
| muscle system | | |
| skeletal system | | |
| nervous system | | |

**3** Complete the table by filling in the main function of each organ system in the body.

# Classification and variation

**Activity 1** **Using and making a key**  Date: _____

Jamilah's dog had eight puppies. Each one was different.

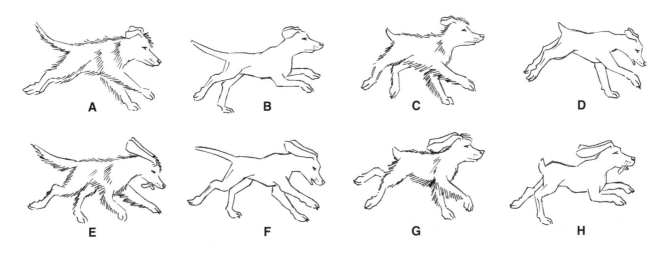

A          B          C          D

E          F          G          H

Jamilah developed this branching key to identify the puppies.

```
                              puppies
              ┌──────────────────┴──────────────────┐
          big ears                              small ears
       ┌──────┴──────┐                      ┌──────┴──────┐
   long tail    short tail              long tail    short tail
   ┌───┴───┐    ┌───┴───┐              ┌───┴───┐    ┌───┴───┐
smooth  furry smooth furry          smooth  furry smooth furry
 coat   coat   coat   coat           coat    coat   coat   coat
  □      □      □      □               □       □      □      □
```

**1** Fill in the letter for each puppy in the boxes at the bottom of the key.

**2** On separate paper, make a numbered key to identify these mice.

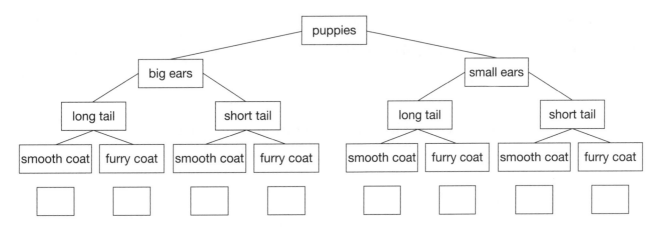

A       B       C       D       E       F       G       H

# Activity 2  Classifying plants

Date:_____

**1** Fill in the missing words in the table.

| Mosses | Ferns |
|---|---|
| These are _____ plants that live in _____ places. They can't survive in hot dry places as they lose _____ through their _____ leaves and can't transport water through the plant. Mosses produce spores, not _____. | Ferns are found naturally in _____, damp places. They are bigger than mosses. Ferns have strong roots, _____ and leaves. The leaves hold _____ and the plants can transport water so they can survive in places that are not damp. Ferns also produce _____, not seeds. |
| **Conifers** | **Flowering plants** |
| These are plants that bear _____. They normally have thin, _____-shaped leaves, which they keep all year. They have a water-transport system and waterproof leaves. The plants produce _____, which are formed inside cones. Conifers can range from small shrubs to _____ trees. | These are plants that produce _____. The flowers produce seeds, inside _____. Flowering plants have a water-transport system and usually have _____ waterproof leaves. Most plants we see fall into this group – from _____ daisies to flowering grasses to massive gum _____. |

**2** Sketch and label one plant from each group.

| Moss | Fern |
|---|---|
| | |
| **Conifer** | **Flowering plant** |
| | |

# Activity 3 Vertebrates and invertebrates

Date:_____

Study the animals in the picture carefully. Use what you can see to group the animals according to their similarities. Write the names of the animals in the correct boxes in the chart on the next page.

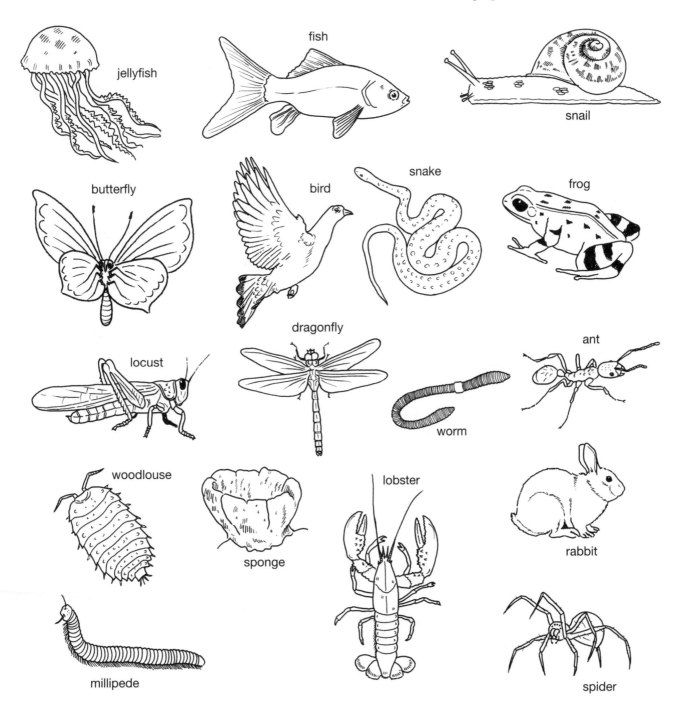

jellyfish

fish

snail

butterfly

bird

snake

frog

locust

dragonfly

worm

ant

woodlouse

sponge

lobster

rabbit

millipede

spider

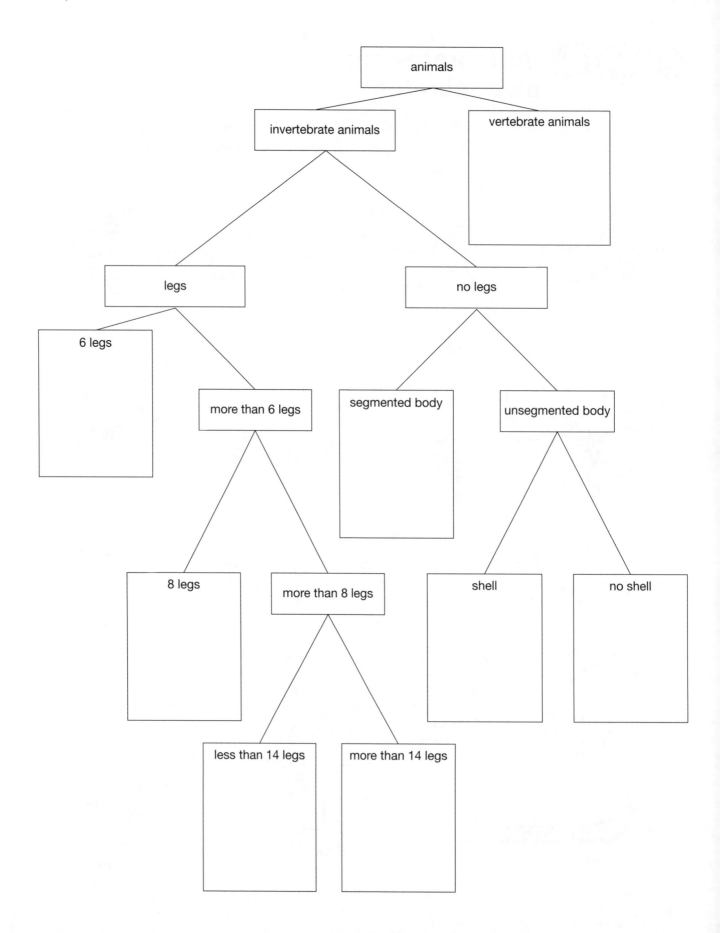

# Activity 4  Classifying vertebrates

Date: _____

Study the animals in the diagram carefully. Use your observations to tick the correct columns in the table on the next page.

perch

owl

iguana

robin

heron

snake

flamingo

parrot

trout

dove

bat

monkey

mackerel

squirrel

frog

cat

dog

lion

tortoise

alligator

shark

| Animal | Hair | No hair | Feathers | No feathers | Fins | No fins | Scales | No scales |
|---|---|---|---|---|---|---|---|---|
| flamingo | | | | | | | | |
| iguana | | | | | | | | |
| trout | | | | | | | | |
| tortoise | | | | | | | | |
| dove | | | | | | | | |
| alligator | | | | | | | | |
| parrot | | | | | | | | |
| heron | | | | | | | | |
| snake | | | | | | | | |
| monkey | | | | | | | | |
| cat | | | | | | | | |
| robin | | | | | | | | |
| dog | | | | | | | | |
| perch | | | | | | | | |
| bat | | | | | | | | |
| frog | | | | | | | | |
| squirrel | | | | | | | | |
| shark | | | | | | | | |
| mackerel | | | | | | | | |
| owl | | | | | | | | |
| lion | | | | | | | | |

## Activity 5  Variation survey

Date:_____

**1** You are going to carry out a survey at school to find out about variation.

You will need to survey at least 20 children.

You are going to observe whether the children:

- have straight or curly hair
- have lobed or unlobed ears
- can roll their tongues or not.

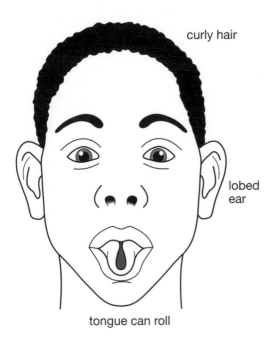

curly hair

lobed ear

tongue can roll

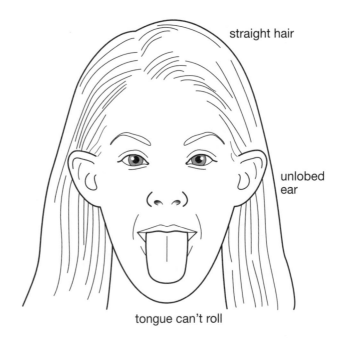

straight hair

unlobed ear

tongue can't roll

**2** Use tally marks to record your results in this table.

| Characteristic | Tally | Total |
|---|---|---|
| curly hair |  |  |
| straight hair |  |  |
| lobed ears |  |  |
| unlobed ears |  |  |
| can roll tongue |  |  |
| cannot roll tongue |  |  |

**3** Draw a bar graph to show your results.

# Understanding ecosystems

**Activity 1** **Matching organisms to their habitats**

Date:_____

**1** Look at the six habitats shown here. Then complete the table on the next page.

coastal habitat

tropical forest

savannah/grassland

pond

dry, rocky desert

city

| Habitat | Conditions found in this habitat | Plants that might be found there | Animals that might be found there |
|---|---|---|---|
| coastal habitat | | | |
| tropical forest | | | |
| savannah/ grassland | | | |
| pond | | | |
| dry, rocky desert | | | |
| city | | | |

**2** Write down four organisms that would *not* be found in any of these habitats. Say why you think each organism could not survive in these conditions.

a) _____

_____

b) _____

_____

c) _____

_____

d) _____

_____

**Activity 2** **How camels are adapted to hot, dry conditions**

Date:_____

**1** Read the information in the box carefully.

> Camels are famous for their humps. Camels store water and fat in their humps so that they can go for ages without drinking water or eating. Camel hair traps a layer of air to keep the animal cool. Their feet are large and flat to support them and make it easier for them to walk on sand.
>
> Camels absorb water vapour from the air through their nostrils. They are also able to close their nostrils to keep sand out. Their long eyelashes help prevent sand getting in their eyes and their ear flaps keep sand out of their ears.

**2** Label the diagram to show how the camel is well adapted to living in the desert.

**3** Polar bears live in the ice and snow in cold polar regions. On separate paper, list some of the adaptations that help them to survive in this habitat. Explain how each one helps.

## Activity 3 · Test your knowledge

Date:_____

**1** In which two habitats would you find termites? Tick *two* boxes.

☐ a river    ☐ sandy soil    ☐ rotting floorboards    ☐ a rocky cliff

**2** How are cactuses adapted to living in hot, dry habitats? Give *two* ideas.

_____

_____

**3** Choose three different environments shown in the picture.
What conditions would you expect to find in each of them?

_____

_____

_____

_____

_____

_____

**4** Refer to the picture and fill in the missing words in these sentences.

**a)** The cold mountain areas (E) are not suitable habitats for _____.

**b)** Fresh water (A) is a good habitat for _____.

**c)** In the sea (B) you are likely to find _____.

**d)** The dark forest (C) is _____ suitable for humans to live in.

**e)** Herds of _____ are likely to be found on the grassy plains (F).

**f)** Hyenas and wild dogs are likely to be found on the _____ as well.

## Activity 4  Observing ecosystems

Date:_____

**1** Observe each animal closely. Complete the table.

| Questions | Snail | Earthworm | Ant | Bee |
|---|---|---|---|---|
| Where does the animal prefer to live? | | | | |
| What conditions does the animal like in its habitat? | | | | |
| How is the animal adapted to its habitat? | | | | |
| What social groups does the animal live in? | | | | |
| My own question: | | | | |

**2** Do this task on separate paper.

   **a)** Draw a large sketch of any large animal that is found in your country.

   **b)** Add labels to show where it lives and how it is adapted to its habitat.

## Activity 5 | Food chains

Date:_____

Joseph and Sandy found these organisms in a vegetable patch.

Sandy said: 'Slugs eat cabbage – I know because I have found some in a cabbage at home. I think the snake would eat the frog.'

Joseph said: 'I think the frog would eat the slug. I agree that the snake would eat the frog.'

**1** Draw a food chain to show the relationships Sandy and Joseph have described.

**2** What is the original source of energy in this food chain?

_____

**3** What is the producer in the food chain you have drawn?

_____

**4** Name *two* consumers in this food chain.

_____

**5** Which organism in the food chain is a herbivore?

_____

**6** Which organisms in the food chain are carnivores?

_____

**7** What would happen in this food chain if the gardener killed all the frogs?

_____

# Chapter 6 Acids and bases

**Identifying acids**

Date:_____

**1** Twelve 'acid' words are hidden in this grid. Find them and shade them using different colours. The first word has been shaded for you.

| A | I | Y | K | G | U | A | C | E | T | I | C | V | H | A |
|---|---|---|---|---|---|---|---|---|---|---|---|---|---|---|
| L | H | Y | D | R | O | C | H | L | O | R | I | C | S | C |
| O | Z | J | B | I | T | I | E | I | D | E | B | A | M | R |
| Q | N | O | X | F | E | D | D | U | S | U | T | R | D | O |
| J | M | J | F | H | N | G | C | H | T | D | P | B | F | S |
| D | L | E | M | O | N | J | U | I | C | E | S | O | E | M |
| P | A | H | A | W | C | Z | G | X | I | G | A | N | L | I |
| H | C | G | L | O | R | F | Y | Q | T | R | M | I | L | K |
| K | T | K | I | N | B | X | C | W | R | L | V | C | B | T |
| R | I | Q | C | F | Z | S | A | F | I | F | I | N | Q | R |
| L | C | M | N | G | N | Y | A | S | C | O | R | B | I | C |
| Q | I | O | W | M | J | E | X | M | U | R | T | D | N | C |
| P | J | K | L | K | Z | W | S | W | J | M | A | Q | U | B |
| K | B | E | U | L | V | D | D | P | V | I | R | O | Z | P |
| V | C | S | U | L | P | H | U | R | I | C | X | T | Y | A |

**2** Where would you find each of these acids?

**a)** formic acid _____

**b)** sulphuric acid _____

**c)** lactic acid _____

**d)** ascorbic acid _____

**e)** tartaric acid _____

**f)** hydrochloric acid _____

**g)** malic acid _____

**h)** carbonic acid _____

**i)** citric acid _____

**Activity 2** **Identifying bases**

Date:_____

**1** Tick the household products on this list that are bases.

☐ soap                      ☐ vinegar
☐ vitamin C                 ☐ oven cleaner
☐ lemon juice               ☐ drain cleaner
☐ floor cleaner             ☐ milk of magnesia
☐ milk                      ☐ water
☐ batteries                 ☐ bicarbonate of soda
☐ flour                     ☐ ammonia
☐ rat poison                ☐ toothpaste
☐ tomato sauce              ☐ caustic soda
☐ bleach                    ☐ aspirin
☐ brass cleaner             ☐ baking powder
☐ cooking oil               ☐ shampoo

**2** Explain the meaning of the following words:

**a)** base

_____

_____

**b)** alkali

_____

_____

## Activity 3 Safety symbols

Date:_____

**1** Complete this table by drawing the missing warning symbols.

| Symbol | Meaning |
|---|---|
| | corrosive – eats away and rots skin and clothing |
| | poisonous – harmful to humans and animals |
| | irritant – burns skin and can affect lungs and eyes |
| 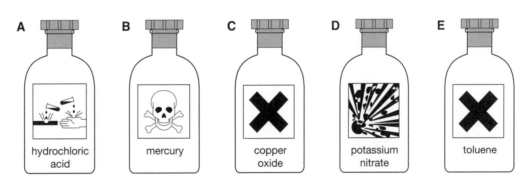 | explosive – can produce loud sound and energy by bursting into pieces |

**2** Look at the chemicals shown here.

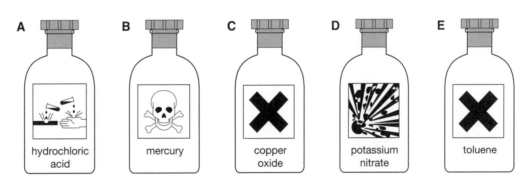

A hydrochloric acid  B mercury  C copper oxide  D potassium nitrate  E toluene

**a)** Which chemical(s) could cause your eyes to water and irritate your lungs? _____

**b)** Which chemical(s) would kill fish if they were spilled in a river?

_____

**c)** Which chemical could destroy a science laboratory if it was not stored properly? _____

**d)** Which chemical would make a hole in your clothes if you spilt a drop of it? _____

**Activity 4** **The pH scale** Date:_____

**1** Colour this scale to show what colours you would see with universal indicator at each pH level.

| 0 |
| 1 |
| 2 |
| 3 |
| 4 |
| 5 |
| 6 |
| 7 |
| 8 |
| 9 |
| 10 |
| 11 |
| 12 |
| 13 |
| 14 |

**2** Add examples of acids at different pH levels down the left-hand side of the scale.

**3** Add examples of bases at different pH levels down the right-hand side of the scale.

**Activity 5** **Test your knowledge**　　　Date:＿＿＿＿＿＿

**1** Write the meaning of each word in the space provided and give an example of each.

**a)** acid ＿＿＿＿＿＿＿＿＿＿＿＿＿＿＿＿＿＿＿＿＿＿＿＿＿＿＿＿＿＿＿＿＿＿＿

＿＿＿＿＿＿＿＿＿＿＿＿＿＿＿＿＿＿＿＿＿＿＿＿＿＿＿＿＿＿＿＿＿＿＿＿＿＿

**b)** alkali ＿＿＿＿＿＿＿＿＿＿＿＿＿＿＿＿＿＿＿＿＿＿＿＿＿＿＿＿＿＿＿＿＿＿

＿＿＿＿＿＿＿＿＿＿＿＿＿＿＿＿＿＿＿＿＿＿＿＿＿＿＿＿＿＿＿＿＿＿＿＿＿＿

**c)** base ＿＿＿＿＿＿＿＿＿＿＿＿＿＿＿＿＿＿＿＿＿＿＿＿＿＿＿＿＿＿＿＿＿＿＿

＿＿＿＿＿＿＿＿＿＿＿＿＿＿＿＿＿＿＿＿＿＿＿＿＿＿＿＿＿＿＿＿＿＿＿＿＿＿

**d)** indicator ＿＿＿＿＿＿＿＿＿＿＿＿＿＿＿＿＿＿＿＿＿＿＿＿＿＿＿＿＿＿＿＿

＿＿＿＿＿＿＿＿＿＿＿＿＿＿＿＿＿＿＿＿＿＿＿＿＿＿＿＿＿＿＿＿＿＿＿＿＿＿

**e)** litmus ＿＿＿＿＿＿＿＿＿＿＿＿＿＿＿＿＿＿＿＿＿＿＿＿＿＿＿＿＿＿＿＿＿＿

＿＿＿＿＿＿＿＿＿＿＿＿＿＿＿＿＿＿＿＿＿＿＿＿＿＿＿＿＿＿＿＿＿＿＿＿＿＿

**f)** pH scale ＿＿＿＿＿＿＿＿＿＿＿＿＿＿＿＿＿＿＿＿＿＿＿＿＿＿＿＿＿＿＿＿＿

＿＿＿＿＿＿＿＿＿＿＿＿＿＿＿＿＿＿＿＿＿＿＿＿＿＿＿＿＿＿＿＿＿＿＿＿＿＿

**g)** neutral ＿＿＿＿＿＿＿＿＿＿＿＿＿＿＿＿＿＿＿＿＿＿＿＿＿＿＿＿＿＿＿＿＿＿

＿＿＿＿＿＿＿＿＿＿＿＿＿＿＿＿＿＿＿＿＿＿＿＿＿＿＿＿＿＿＿＿＿＿＿＿＿＿

**2** Complete this diagram by filling in the missing information in the boxes.

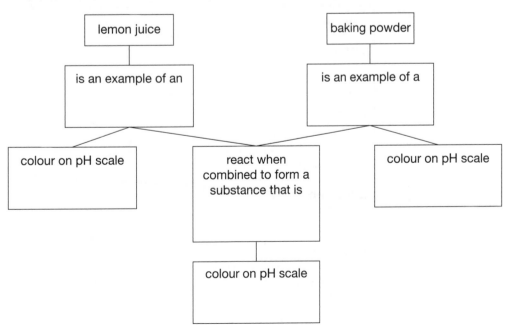

# Chapter 7  Physical and chemical changes

**Activity 1**  **Summarising what you have learned**       Date:_____

**1** Tick the boxes for solids, liquids and gases to show which statements apply to each phase of matter.

| Solids | Liquids | Gases | Characteristics |
|--------|---------|-------|-----------------|
|        |         |       | have a fixed shape |
|        |         |       | take the shape of the container they are in |
|        |         |       | spread out to fill all the available space |
|        |         |       | have a mass that can be measured |
|        |         |       | cannot be compressed |
|        |         |       | can be compressed easily |
|        |         |       | can change from one state to another |

**2** Tick the correct columns below to show which state each substance is in at room temperature.

| Substance | Solid at room temperature | Liquid at room temperature | Gas at room temperature |
|-----------|---------------------------|----------------------------|-------------------------|
| oxygen |  |  |  |
| water |  |  |  |
| ice |  |  |  |
| petrol |  |  |  |
| iron |  |  |  |
| candle wax |  |  |  |
| ice-cream |  |  |  |
| butter |  |  |  |
| chocolate |  |  |  |
| your breath |  |  |  |

**Activity 2** **Labelling diagrams** Date:_____

**1** A group of pupils did the experiment below to show how water changes its state in the water cycle. Use the words in the box underneath to correctly label the diagram.

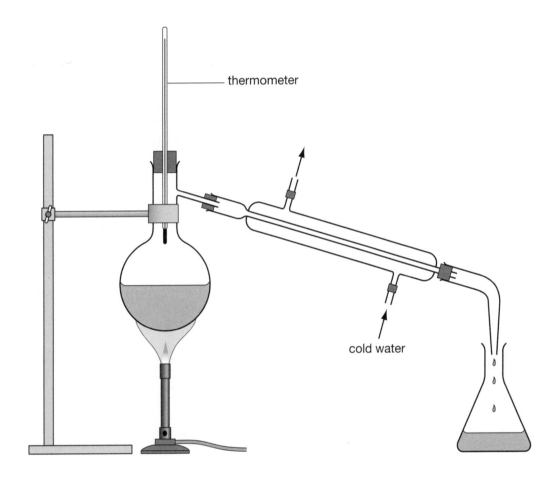

thermometer

cold water

| heating | evaporation | water | water vapour | cooling | condensation | water droplets | water |

**2** Use the headings below to write up a report of this experiment, on separate paper.

**Aim**

**Equipment**

**Method**

**Observations**

**Conclusions**

## Activity 3  Explaining changes of state    Date:_____

**1** Where does water go to when it boils away?

_____

**2** Where does sugar go when you stir it into hot water?

_____

**3** Where do clouds come from?

_____

**4** Why does water form on the outside of a can of cold drink when you take it out of the refrigerator?

_____

**5** Why can you see your breath on a cold day?

_____

_____

**6** How is rain formed?

_____

_____

**7** Where would you put a bird bath in your garden to stop the water from evaporating too quickly? Give reasons for your answer.

_____

_____

_____

_____

**8** Your friend wants to know if all liquids evaporate at the same speed. Explain how you would do a fair test to find the answer to the question.

_____

_____

_____

_____

**Activity 4** **Measuring temperature** Date:_____

**1** Write down the temperature shown on each of these Celsius thermometers.

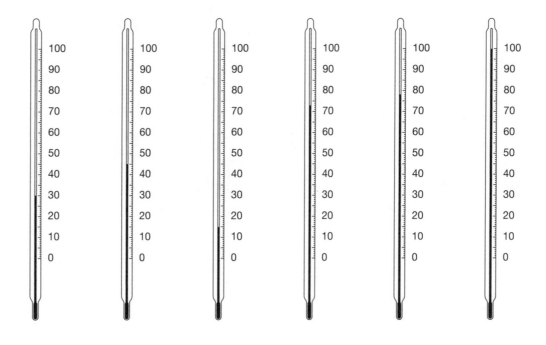

_____   _____   _____   _____   _____   _____

**2** Draw a red line on each thermometer to show the temperature given below it.

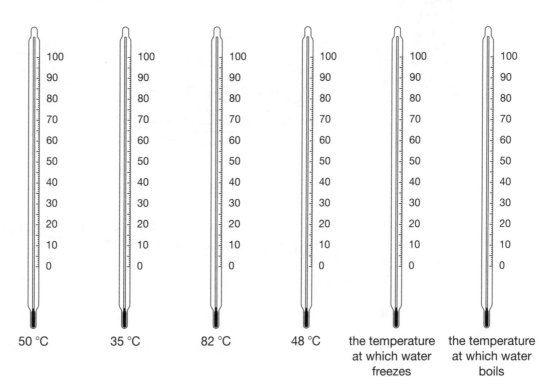

50 °C       35 °C       82 °C       48 °C    the temperature at which water freezes    the temperature at which water boils

## Activity 5 | Identifying changes

Date:_____

Look at the changes shown in each diagram.

_____

_____

_____

_____

_____

_____

**1** Write the correct scientific word under each picture to describe the change that it shows.

**2** In red, circle the changes that are caused by heating.

**3** In blue, circle the changes that are caused by cooling.

**4** In green, underline all the physical (reversible) changes.

# Activity 6 Crossword puzzle

Date: _____

Use the clues to complete this crossword puzzle.

## Clues

### Across

**1** Matter that has a fixed shape.

**2** When you _____ butter, it melts.

**3** Water in a solid state.

**4** Process that can change candle wax into carbon and water.

**5** Type of change that can be reversed easily.

**6** What water does when it changes to ice.

**7** The name of the change from a gas back into a liquid.

### Down

**a** The name of the change from a gas to a solid without becoming a liquid.

**b** Milk is a _____ because it takes the shape of its container.

**c** Type of change that cannot easily be reversed.

**d** Water freezes at _____ degrees Celsius.

**e** When water is left in the sunshine, it will _____.

**f** Air is an example of matter in this state.

**g** If you heat chocolate, it will _____ from a solid to a liquid.

# Materials and their properties

**Activity 1** | **Choosing the right materials**    Date:_____

Study the objects in the table carefully. The table continues on the next page.

Write down why the materials used to make them are not suitable.

Suggest *two* materials that are more suitable for making each object.

| Object | Why this material is not suitable | Two materials that would be more suitable |
|---|---|---|
| metal shoes | | |
| rubber tennis racket | | |
| glass car | | |
| plastic cooking pots | | |

| Object | Why this material is not suitable | Two materials that would be more suitable |
|---|---|---|
| cardboard nails | | |
| wooden envelope | | |
| thin fabric suitcase | | |
| rock curtains | | |
| rubber coins | | |
| paper balloons | | |

**Activity 2** **Identifying strong materials** Date:_____

Walk around your school and observe the materials that have been used to make different things.

Give *three* examples of uses of each of the strong materials in the table below.

| Material | Example 1 | Example 2 | Example 3 |
|---|---|---|---|
| strong wood | | | |
| strong metal | | | |
| strong plastic | | | |
| strong textiles | | | |
| strong glass | | | |
| strong ceramics | | | |

## Activity 3 | **Flexible materials**

Date:_____

**1** Find *ten* flexible objects. Write down their names on rough paper.

**2** Sort the objects in your list in order of how flexible they are. Write the most flexible object first and the least flexible object last.

| most flexible | |
|---|---|
| | |
| | |
| | |
| | |
| | |
| | |
| | |
| | |
| least flexible | |

**3** Find out what each object is made from. Then complete these sentences.

**a)** Many flexible objects are made from _____.

**b)** Objects made from _____ are not very flexible.

**c)** Very few flexible objects are made from _____.

**Activity 4** **Metals and their properties**   Date:_____

**1** Fill in the missing words in the sentences below.

  **a)** Metals are solids that have a bright _____ surface when cut.

  **b)** Metals are generally _____ – they can be formed into shapes.

  **c)** Ductile means it can be pulled into thin _____.

  **d)** Metals are _____ conductors of electricity.

**2** Give *three* metals that are used for expensive jewellery.

_____

_____

_____

**3** Fill in the boxes to show what materials you would use to make each part of the bicycle. Give reasons for your choices.

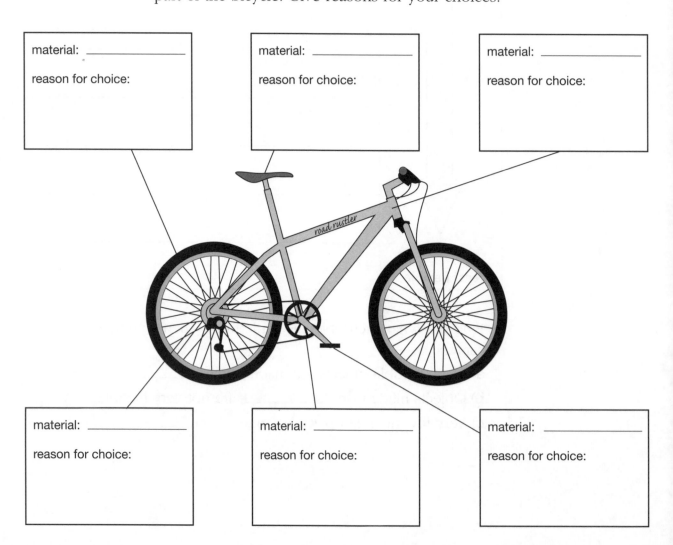

material: _____

reason for choice:

material: _____

reason for choice:

material: _____

reason for choice:

material: _____

reason for choice:

material: _____

reason for choice:

material: _____

reason for choice:

**Activity 5** **Choosing the best metal** Date:_____

When people use metals to make things, they often mix different metals together. A metal mixed with another metal is called an alloy. The table gives you information about different types of alloys and their properties.

| Type of alloy | Properties |
|---|---|
| brass | corrosion resistant, strong |
| solder | melts easily at fairly low temperatures |
| duriron | acid resistant |
| cupronickel | elastic, corrosion resistant |
| invar | does not expand or contract when heated or cooled |
| permalloy | magnetic |
| stainless steel | strong, corrosion resistant |
| chrome–vanadium | strong, stain resistant |
| tungsten steel | stays hard at high temperatures |

**1** Use the information in the table to choose an alloy for each purpose in the table below.

| Purpose | Alloy chosen | Reason for choice |
|---|---|---|
| ball bearings, tools, axles of cars | | |
| gears and drive shafts | | |
| bridge and other support cables | | |
| cutting tools, drill bits, saw blades | | |
| pipes, containers | | |
| railway tracks, armour plating | | |
| surgical instruments, knives | | |
| magnets | | |
| electrical cables | | |

**2** Complete the table by giving a reason for each choice.

The particle model

**Explaining the particle model**

Date: _____

**1** The diagrams below show how the molecules are arranged in solids, liquids and gases. Label them and tell a partner why they look different from each other.

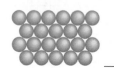 _____  _____ 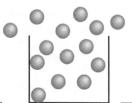 _____

**2** Explain what happens when:
   **a)** you heat a liquid

   _____

   **b)** you heat a solid

   _____

   **c)** you cool down a gas.

   _____

**3** Explain why:
   **a)** some bridges are built with rubber rollers at the ends

   _____

   **b)** an electric kettle switches itself off when the water boils

   _____

   _____

   **c)** you can heat up the lid of a jar to make it easier to open

   _____

   _____

   **d)** ice-cream melts if you leave it out of the freezer.

   _____

   _____

**Activity 2** **What happens when you heat a gas?**

**1** Write labels to explain what is happening in this investigation.

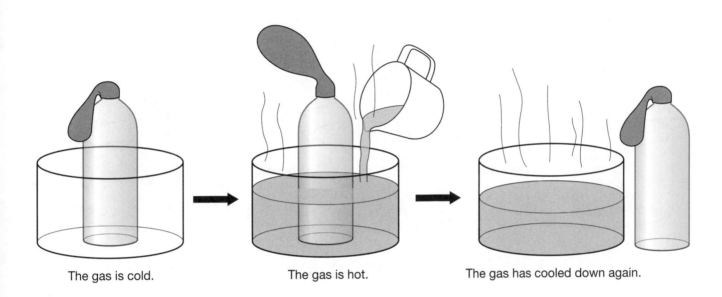

The gas is cold.                    The gas is hot.                    The gas has cooled down again.

**2** Fill in the word 'more' or the word 'less' in the spaces below, to make each sentence true.

**a)** When gas molecules are heated, they need _____ space to move about.

**b)** The gas will take up _____ space.

**c)** When gas molecules cool down they need _____ space to move about.

**d)** The gas will then take up _____ space.

**3** Complete these sentences using correct scientific words.

**a)** When a gas is heated it _____.

**b)** When a gas is cooled it _____.

# Activity 3 What happens when you heat a liquid?

Date:_____

The diagram shows what two pupils did to investigate what happens when you heat a liquid.

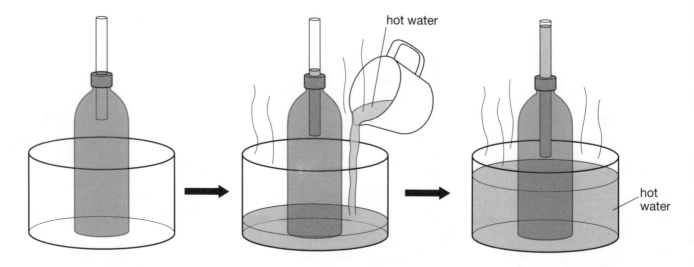

**1** Use the headings below to write up this investigation, on separate paper.

**Aim**

**Equipment**

**Method**

**Observations**

**Conclusions**

**2** In your conclusion, copy and complete these sentences.

**a)** When water is heated it _____.

**b)** When water is cooled it _____.

# Activity 4   Scientific words

Date: _____

1  Find all the key words from Chapter 9 in your coursebook.
Write them in the first column of the table.

2  Use the glossary in your coursebook (pages 175 to 178) and a
dictionary to complete the table.

| Keyword | Meaning in science | Meaning in everyday life |
|---|---|---|
|  |  |  |
|  |  |  |
|  |  |  |
|  |  |  |
|  |  |  |
|  |  |  |
|  |  |  |
|  |  |  |
|  |  |  |
|  |  |  |
|  |  |  |
|  |  |  |
|  |  |  |
|  |  |  |

**Activity 5** **Test your knowledge**          Date:_____

Fill in the missing words in these sentences.

**1** Matter is made up of small particles called _____.

**2** Molecules move faster and move _____ _____ when they are heated.

**3** Molecules slow down and move _____ _____ when they are cooled.

**4** When water changes to a _____, evaporation has taken place.

**5** When ice _____, it changes from a solid to a liquid.

**6** _____ is caused by particles in matter pressing down, or pressing against the sides of the container they are in.

**7** _____ cannot be compressed, _____ can be slightly compressed if they are cooled and _____ can be compressed easily.

**8** Engineers have to allow for _____ and _____ when they build roads and bridges.

**9** When solids _____ in water, they form a solution.

**10** We can smell perfume and other smells in the air because of _____.

# Mixing and separating substances

**Activity 1** **Making mixtures**                     Date:_____

Draw or describe *two* examples of each type of mixture below.

| Type of mixture | Examples |
|---|---|
| solid-with-solid | |
| liquid-with-liquid | |
| gas-with-gas | |
| solid-with-liquid | |
| solid-with-gas | |
| liquid-with-gas | |

**Activity 2** **Soluble or insoluble?**       Date:_____

**1** Sort the materials in the box according to whether they are soluble in water or insoluble in water.

| bath salts     sand     flour     tea     rice     coffee powder     salt |
| coffee grounds     talcum powder     sugar     starch     fertiliser |

| Soluble in water | Insoluble in water |
|---|---|
|  |  |

**2** Name the solvent and the solute in each solution below.

**a)** hot black coffee

solvent: _____

solute: _____

**b)** chocolate milk

solvent: _____

solute: _____

**c)** soapy washing-up water

solvent: _____

solute: _____

**d)** garlic oil

solvent: _____

solute: _____

**e)** liquid plant food

solvent: _____

solute: _____

**Activity 3** **Saturated solutions**                          Date:_____

You know that salt and sugar can dissolve in water.

Is there a limit to how much salt or sugar can dissolve in 100 millilitres of water?

**1** On separate paper, write down an investigation you could do to find this out. Use these headings.

**Aim**

**My hypothesis**

**Equipment**

**Method**

**2** Predict what you would observe if you did this investigation.

**3** What would you need to do to make sure the investigation was a fair test?

**4** What methods would you use to record and present your results? Why?

## Activity 4  Separating mixtures

Date:_____

The table shows different methods of separating mixtures.

Write down each of your examples of mixtures from Activity 1 next to the method (or methods) you could use to separate it.

| Method of separation | Examples of mixtures that can be separated by this method |
|---|---|
| sorting by size | |
| sifting | |
| decanting | |
| filtering | |
| magnetic separation | |
| distillation | |
| chromatography | |

## Activity 5  Science in action

Date:_____

**1** This photograph shows a salt factory.

Write a few sentences explaining how this factory separates salt from sea water.

_____

_____

_____

_____

_____

_____

_____

_____

_____

**2** This is a diagram of a filter bed used to remove solid waste from water in a large treatment plant.

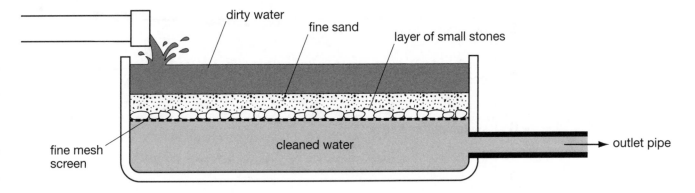

dirty water · fine sand · layer of small stones · fine mesh screen · cleaned water · outlet pipe

**a)** How does the filter bed help to separate the dirt from the water?

_____

_____

_____

**b)** Explain why the water that comes from the filter bed will still not be clean enough to drink.

_____

_____

# Chapter 11 Forces

**Activity 1** **Naming forces**          Date:_____

Complete the table by filling in the type of force for each action in rows 1 to 8, and by filling in an example of each type of force in rows 9 to 11.

| | Action | Type of force (push, pull or twist) |
|---|---|---|
| 1 | picking up a parcel | |
| 2 | hammering a nail into wood | |
| 3 | screwing a nail into wood | |
| 4 | kicking a ball | |
| 5 | towing a boat behind a car | |
| 6 | opening a door | |
| 7 | pedalling a bicycle | |
| 8 | switching on a light | |
| 9 | | twist |
| 10 | | pull |
| 11 | | push |

# Activity 2 Using arrows to show forces

Date:_____

Draw arrows to show the forces in each situation.

**Activity 3** **Correcting false statements** Date:_____

The following statements about forces are all false.

Rewrite each statement so that it is true.

**1** In a non-contact force, the object causing the force must touch the object or material that is feeling the force.

_____

_____

**2** In a non-contact force, the objects touch each other.

_____

**3** A magnet can pull metal objects towards it without touching them. This is a good example of a contact force.

_____

_____

**4** A force cannot make an object move.

_____

**5** A force cannot change the speed of an object.

_____

**6** A pulling force can compress (squash) some things (such as bread dough or a spring).

_____

_____

**7** A very soft pull will stretch an elastic band more than a hard pull will.

_____

_____

**8** Forces are measured in units called kilograms (kg).

_____

# Activity 4 · Reading graphs

Date: _____

The graph shows the force that was needed to stretch five different springs to twice their original length.

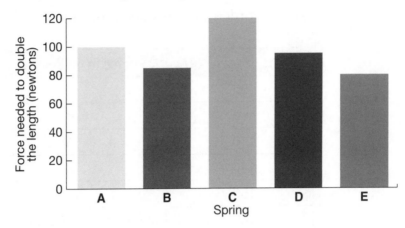

**1** Which spring do you think was the weakest? Why?

_____

**2** Which spring do you think was the strongest? Why?

_____

**3** Which spring needed the largest force to stretch it to twice its length?

_____

**4** Which spring needed the smallest force to stretch it to twice its length?

_____

**5** Which spring needed a force of 90 newtons to stretch it to twice its length?

_____

**6** Which spring is likely to have been made from the thinnest wire? Why?

_____

**7** Which spring is likely to have been made from the thickest wire? Why?

_____

**8** If a weight of 50 newtons was hung from each spring, which one would be likely to stretch the most? Give a reason for your answer.

_____

_____

_____

## Activity 5    Drawing graphs

Date:_____

Two pupils hung mass pieces from a small spring to see how far it would stretch.

They recorded their results in this table.

| Mass in grams (g) | Stretch in millimetres (mm) |
|---|---|
| 50 | 2 |
| 80 | 4 |
| 110 | 6 |
| 140 | 8 |
| 150 | 10 |
| 180 | spring stretched out and would not spring back |

**1** Draw a graph to show this data.

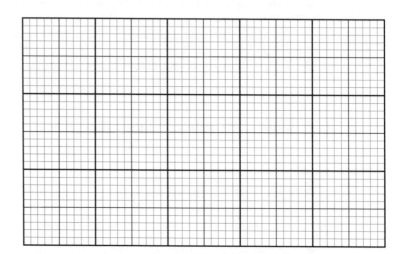

**2** What patterns can you see in your graph?

_____

_____

_____

_____

# Chapter 12  Energy resources

**Matching statements to pictures**

Date:_____

Study these pictures carefully. Then carry out the activity on the next page.

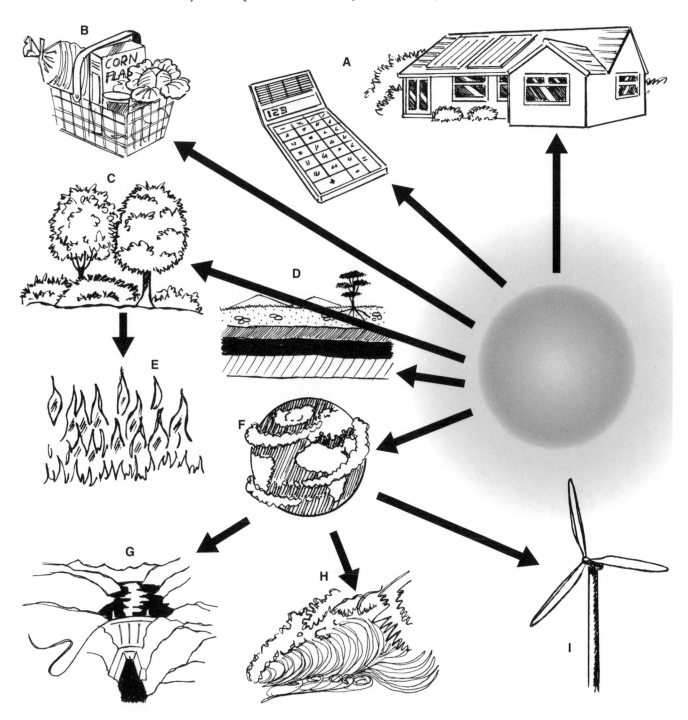

Choose the picture (A to I) that matches each statement. Write the letter next to the statement.

_____ Solar panels can absorb and store the Sun's energy.

_____ The Earth's weather is caused by the heat from the Sun.

_____ Plants use the Sun's energy to make their own food.

_____ Rain is stored in lakes and dams. The stored water can be used to make electricity.

_____ When plants (wood) are burned, they release energy.

_____ Waves have energy that can be used to make electricity.

_____ Animals eat plants to get their energy. Humans eat food that comes from plants and animals to get their energy.

_____ The remains of plants and animals that died millions of years ago have formed fossil fuels like coal and oil.

_____ Wind energy can be used to pump water and supply power.

**Activity 2** **Energy flow diagrams**

Date:_____

**1** Study the pictures carefully.

A    B    C    D

E    F    G    H

**2** In each picture, colour in the things that are moving.

**3** Choose five more examples from the pictures and complete flow diagrams to show where the energy comes from for each movement.

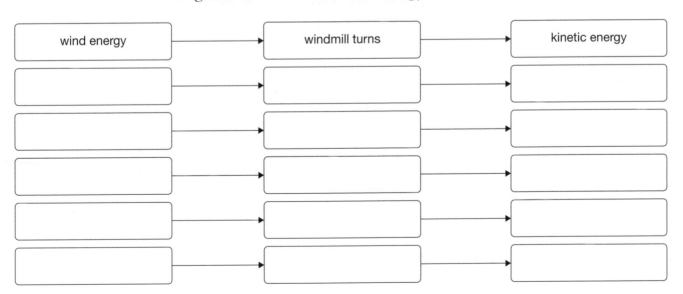

| wind energy | → | windmill turns | → | kinetic energy |

## Activity 3 Fuels

Date:_____

**1** There are eight fuels hidden in the grid. Find them and write down their names. The first one has been shaded for you.

| O | I | L | A | C | O | A | L | P | E |
|---|---|---|---|---|---|---|---|---|---|
| F | A | S | B | F | R | Y | Y | E | D |
| D | I | E | S | E | L | P | U | T | G |
| E | S | R | Y | H | U | J | N | R | D |
| A | X | R | B | V | S | D | R | O | Y |
| T | C | H | A | R | C | O | A | L | W |
| G | Q | W | E | R | T | H | Y | G | O |
| A | U | I | O | P | Z | R | W | H | O |
| S | P | A | R | A | F | F | I | N | D |

_____

_____

_____

_____

_____

_____

_____

**2** Study the graph carefully. It shows how much energy it takes to travel 3 kilometres using different types of transport.

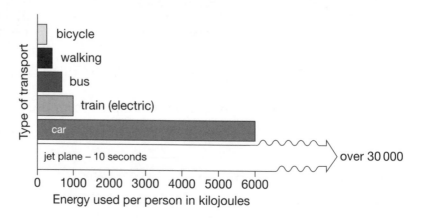

**a)** Which type of transport uses least energy? _____

**b)** Which type of transport uses most energy? _____

**c)** What type of fuel would be used for each type of transport?

_____

**d)** Why does cycling use less energy than walking?

_____

**e)** Which wastes most energy – a bus or a car? Give a reason for your answer.

_____

**Activity 4** **Transferring heat energy**    Date:_____

**1** Look at the cups in the photograph.

**a)** What material was used to make each cup?

_____

_____

**b)** Give an advantage and a disadvantage of using each material to make a cup.

_____

_____

_____

_____

**2** Draw a labelled diagram to show how convection currents allow an electric heater to heat up a whole room.

**Activity 5** **Are we running out of fuel?** Date:_____

Read this newspaper article carefully.

# Who is using all the fuel?

Half of the world's people make use of wood as a fuel for their daily requirements. They burn it for heating, cooking and light. All over Africa and Asia, trees are being chopped down and people, mostly women and girls, are having to walk greater and greater distances in search of fuel.

The fact that firewood is becoming more scarce might suggest that these people are using up the world's fuels. But the truth is that wood makes up only about 5% of all the fuel energy in the world today. This means that the half of the world's people who live in the poorer countries use 5% of the world's fuel. The other half, who live mainly in the industrialised countries of Europe and America, use 95%! Most of this 95% comes from fossil fuels.

The question is: how soon will the fossil fuels run out?

Experts estimate that if we continue to use these fuels at the present rates, we are likely to run out of oil in the next 50 years and natural gas in the next 100 years. Coal supplies will last longer, but mining will have to take place in remote and protected environments.

**1** Which countries use most of the world's energy?

_____

**2** Why do you think these countries use so much energy?

_____

**3** What is the main source of fuel in Africa and Asia? _____

**4** How could people prevent shortages of wood?

_____

**5 a)** Look at the graph. Are people using less or more fuel now than in the past?

_____

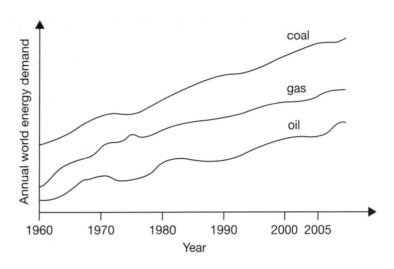

**b)** How do you know this?

_____

_____

**c)** What does this graph suggest about fossil fuel supplies in the future?

_____

_____

Electrical circuits

**Using electrical symbols**          Date:_____

1 Draw a line to match each component to the correct circuit symbol.

2 Write the name of each component in the space next to its symbol.

## Activity 2   Circuit diagrams

Date:_____

**1** List the components used to build each of these circuits.

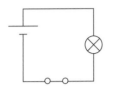 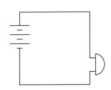 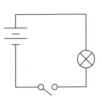 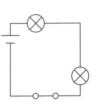 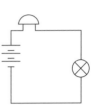

_____   _____   _____   _____   _____

_____   _____   _____   _____   _____

_____   _____   _____   _____   _____

**2** Draw circuit diagrams to show these circuits.

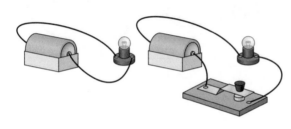 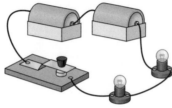 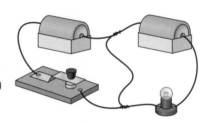

# Activity 3 · Circuit problems

Date: _____

These are the circuits that some pupils built.

None of the lights will work.

Write a reason why the lights will not work in each circuit.

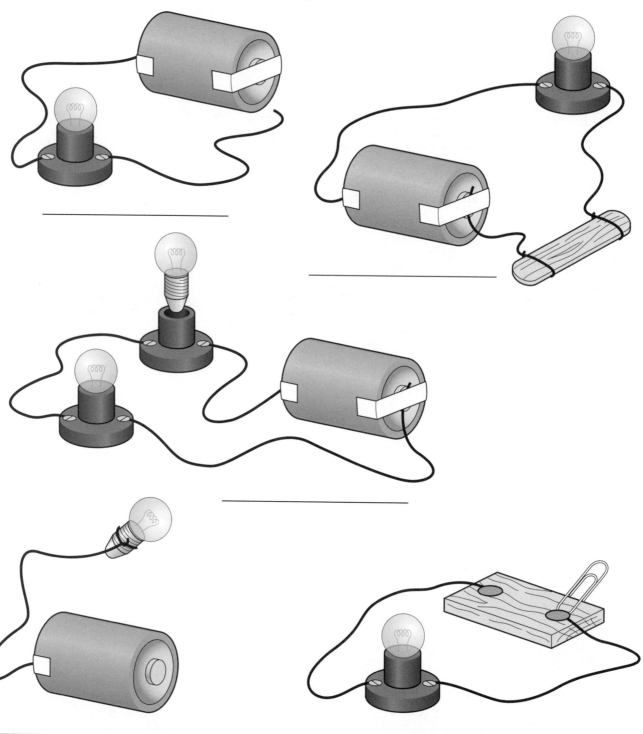

_____

_____

_____

_____

_____        _____

## Activity 4 Evaluating switches

Date: _____

**1** Study these three switches carefully. Try to see how each one was made and how it works.

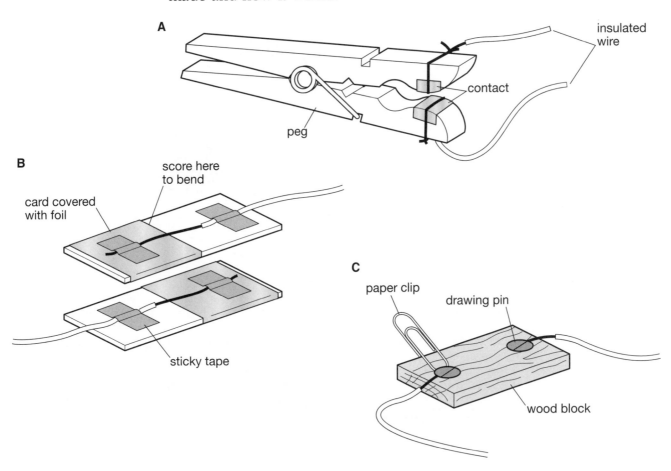

A

insulated wire

contact

peg

B

score here to bend

card covered with foil

sticky tape

C

paper clip

drawing pin

wood block

**2** Complete this table based on your observations.

| Switch | Materials used | How it works |
|--------|----------------|--------------|
| A | | |
| B | | |
| C | | |

**3** Which switch do you think is best? _____

Why? _____

_____

**Activity 5** **Conductors and insulators**     Date:_____

Four pupils made this test circuit to find out whether various materials were conductors or insulators of electricity.

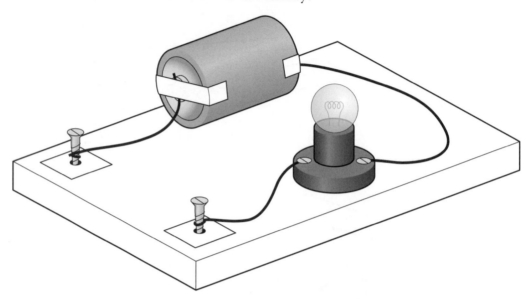

They placed each item in the circuit to see whether the bulb lit up or not.

Use your own knowledge of testing materials to complete this table.

| Material tested | What happened to the bulb? | Conductor | Insulator |
|---|---|---|---|
| nail | | | |
| paper clip | | | |
| wooden stick | | | |
| shoe lace | | | |
| glass rod | | | |
| plastic spoon | | | |
| plastic pen | | | |
| metal pen | | | |
| pair of compasses | | | |
| lemon | | | |

## Activity 6  Test your knowledge

Date:_____

**1** Complete these sentences:

When you _____ the switch, you complete the _____.

_____ can then flow around the circuit. The _____ lights

up. When you _____ the switch, the wires are separated and the

circuit becomes _____. Electricity cannot _____ around the

circuit and the bulb goes _____.

**2** Write a sentence to show what each of these words means.

**a)** electrical circuit

_____

**b)** insulator

_____

**c)** resistor

_____

**d)** conductor

_____

**e)** electrical switch

_____

**3** Circle the insulators in this list.

copper wire          steel

glass                plastic

wood                 rubber

metal spoon          paper clip

water

# Chapter 14   The Earth in space

## Activity 1   Wordsearch

Date:_____

There are 18 words about space in this grid. Find the words and circle them.

Write the words below the box and make sure you know what they all mean.

| A | M | P | Z | E | A | M | E | R | C | U | R | Y | H | U |
|---|---|---|---|---|---|---|---|---|---|---|---|---|---|---|
| G | A | X | O | G | E | O | D | V | K | Q | P | W | C | R |
| W | R | A | N | N | A | O | L | D | O | F | H | J | I | A |
| B | S | A | T | U | R | N | O | V | B | R | K | S | U | N |
| E | X | H | D | C | T | C | N | E | P | T | U | N | E | U |
| C | Z | R | G | F | H | I | J | N | T | L | U | Q | P | S |
| L | P | Y | N | E | F | S | C | U | T | S | S | R | K | B |
| I | E | Q | E | T | E | L | E | S | C | O | P | E | N | E |
| P | D | C | K | S | O | Z | I | S | V | Y | A | M | G | L |
| S | H | P | F | N | W | V | S | L | B | U | C | O | O | L |
| E | M | L | T | L | S | X | U | M | Y | X | E | Z | R | I |
| R | J | U | P | I | T | E | R | T | S | L | A | H | B | P |
| P | D | T | Q | X | A | M | E | V | B | D | C | Y | I | S |
| U | N | O | B | J | R | F | T | W | A | Z | K | V | T | E |
| W | A | T | M | O | S | P | H | E | R | E | M | G | I | A |

_____   _____   _____

_____   _____   _____

_____   _____   _____

_____   _____   _____

_____   _____   _____

_____   _____   _____

# Activity 2 | Labelling diagrams

Date: _____

Write the names of the planets on this diagram of the Solar System.

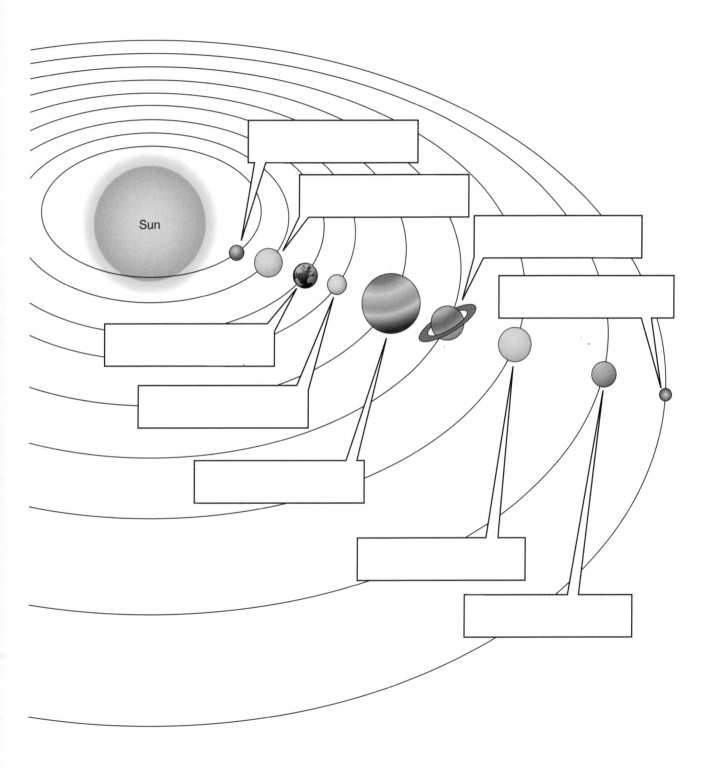

## Activity 3 Drawing an ellipse

Date:_____

The orbits of planets are in the shape of an ellipse, or flattened oval.
Follow the instructions in the diagram to draw an ellipse.

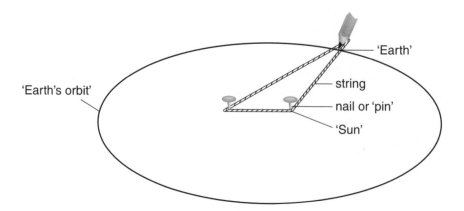

'Earth's orbit'

'Earth'

string

nail or 'pin'

'Sun'

## Activity 4   The Moon

Date: _____

Draw the shapes that the Moon would appear at different times during this month.

| 30 ◐ First quarter | 31 | 1 | 2 | 3 | 4 | 5 |
|---|---|---|---|---|---|---|
| 6 | 7 ○ Full Moon | 8 | 9 | 10 | 11 | 12 ◗ |
| 13 | 14 ◐ Last quarter | 15 | 16 | 17 ◖ | 18 | 19 |
| 20 | 21 ● New Moon | 22 | 23 | 24 | 25 | 26 |
| 27 | 28 | 29 ◐ First quarter | 30 | 31 | 1 | 2 |

## Activity 5  Eclipses

Date:_____

**1** Draw and label the position of the Earth and the Moon in relation to the Sun during a lunar eclipse.

**2** Draw and label the position of the Earth and the Moon in relation to the Sun during a solar eclipse.

**Activity 6** # Test your knowledge

Date:_____

For questions 1–3, draw a ring around the correct answer.

**1** The first person to see that the Earth was round was:

  **a)** Ptolemy, an ancient astronomer

  **b)** Yuri Gagarin, a Russian cosmonaut who went into orbit in 1961

  **c)** Neil Armstrong, an American astronaut who landed on the Moon in 1969

  **d)** Galileo, an astronomer in 1609.

**2** Weather is experienced in only one layer of the atmosphere. This layer is called the:

  **a)** ozone layer

  **b)** lithosphere

  **c)** stratosphere

  **d)** troposphere.

**3** The gas which makes up the biggest percentage of our air is:

  **a)** oxygen

  **b)** carbon dioxide

  **c)** nitrogen

  **d)** hydrogen.

**4** Write down the name of the planet that best matches the features given below:

  **a)** biggest _____

  **b)** travels furthest from the Sun in its orbit _____

  **c)** no atmosphere _____

  **d)** looks blue when seen from Earth _____

  **e)** closest to the Sun _____

  **f)** can sustain human life _____

**5** On separate paper, draw and label diagrams showing the positions of the Earth, Sun and Moon that would result in:

  **a)** a solar eclipse

  **b)** a lunar eclipse

  **c)** a Full Moon.